An Introduction to Depyrogenation Tunnels and Sterile Barrier Systems

Brendan Cooper

Contents

Depyrogenation Tunnels

What is Depyrogenation?

Depyrogenation is a thermal process that involves the removal of pyrogens from components (e.g. vials or containers) that are used for injectable pharmaceuticals and biopharmaceuticals. A pyrogen is defined as any substance that can cause a fever. Bacterial pyrogens include endotoxins. **Later on we shall see that endotoxins are used to challenge Depyrogenation tunnels**. Depyrogenation tunnel design varies depending on the manufacturer; however, they usually consist of the following components.

- ➤ Infeed and Preheating
- ➤ Heating Zone
- ➤ Cooling Zone
- ➤ Outfeed and transport to next unit operation
- ➤ Automatic Emptying

Endotoxins and Depyrogenation

Endotoxins are used to challenge the effectiveness and consistency of depyrogenation tunnels. Endotoxin challenge vials must processed through a Depyrogenation process must demonstrate a ≥ 3 log reduction in endotoxin. Typically, endotoxin challenge vials are placed in close proximity to thermocouples. Using this approach, the temperature profile of the position can be obtained during a cycle. Endotoxin challenge testing is often done during SAT and process development. It is also a requirement of Validation, however, no commercial product can be used during a Depyrogenation tunnel performance qualifications using Endotoxins, as the product would be potentially contaminated with endotoxins. Therefore, Performance validation of Depyrogenation processes results in the discarding of the vials or ampules.

Depyrogenation Tunnels generate tremendous amount of heat and can operate up to temperatures as high as 320°C (depending on design and operational constraints). Air handling is also a key function of the Depyrogenation tunnel. Tunnels should not allow non-sterile air from the room into the sterile air inside tunnel zones.

This is done through the use of HEPA filters and an overpressure cascade approach of the tunnel compared to the surrounding room or environment. Air flow must be laminar in nature to ensure the tunnel can maintain the correct pressures and temperatures. Most tunnels are divided into two sections:

-Hot Zone (Depyrogenation)
-"Cool Zone (Sterilisation /Cooling)

The Depyrogenation section typically operates at higher temperatures, in excess of 270°C which is the recognised Depyrogenation temperature. Depending on the technical specification of the components set-points of 290°C, 300°C or 320°C can be used. Components move slowly through the Depyrogenation stages of tunnels. The "Sterilising Cooling Section" operation mode sterilises the cooling sections. Sterilisation cycle consists of the following steps:

1. Pressure drop
2. Draining heat exchanger
3. Heating up of cooling sections set value of temperature (e.g. 240°C)
4. Sterilisation cooling sections: keeping temperature at recipe set value for recipe set value of time
5. Cooling down without heat exchanger: until temperature < 95°C
6. Cooling down with heat exchanger: until temperature < 25°C

The key requirement of Coolzone Sterilization is that the temperature within the zone is maintained at a minimum of 170°C for a period of no less than 2 hours. This gives a very high degree of assurance that the zone is sterile and suitable for sterile manufacturing operations to occur. In summary, the endotoxin challenge must be sufficient to demonstrate a ≥3 log reduction in endotoxin.

Biological Indicators for Dry Heat

Biological Indicators (BIs) (most commonly Bacillus atrophaeus) are used to demonstrate the efficacy of cool zone sterilization in Depyrogenation tunnels. Using a known indicator population and D-value the the delivered lethality needed to obtain a SAL of at least 10^{-6} can be determined.

The lethality of a cycle can be calculated using the below equation.

Lethality, $F(h) = \Delta T \times \Sigma L$

$L = 10(t\text{-}to) / Z$

$Z = 20$ constant

$to = 170$ the base temperature ($^\circ$C)

$t =$ actual temperature ($^\circ$C)

$\Sigma L =$ cumulative sum of time

$\Delta T =$ Time differential (scan time)

Isolator Barrier Systems

An isolator is a complex barrier system designed to support aseptic processing and manufacturing. The supplied air to such system is generally supplied through a microbially retentive filtration system. High efficiency particulate air (HEPA) filters which are capable of removing particles as small as 0.3μm. Hence, HEPA filters are an integral part of Isolator technology.

Figure : Photograph of a typical Isolator, showing isolator doors fillted with glove ports.

HEPA filters should be capable of achieving Grade A (ISO Class 4.8) at rest and in operation. Some exceptions are permitted, such as powder filling, however, risk assessments should mitigate risk to patients. The Isolator is a sealed enclosure where there is no direct opening to the external environment or room. Transfer of materials or utensils is done in a controlled manner using and a decontaminated interface.

Isolator Interfaces

Depending on the design considerations and individual Vendor designs, Isolators can have a number of operation interfaces. The term "interface" refers to the ability of an operator or process technician to interact with the machine. The primary method of intervention utilizes Glove systems.

Four part Glove systems consisting of a gauntlet, glove, cuff-ring and sleeve. When used properly and by trained personnel, glove systems support critical line interventions required during aseptic processing and manufacturing.

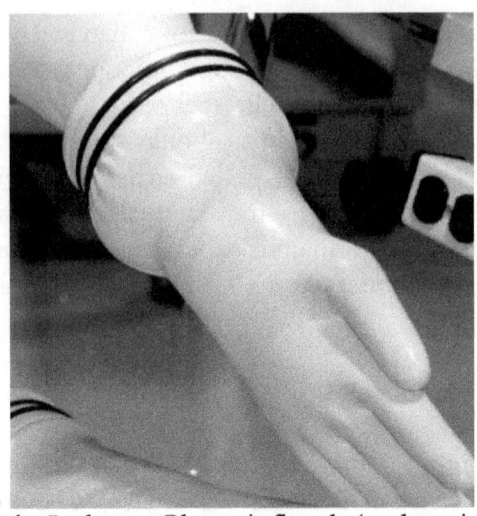

Figure 1: Isolator Glove inflated (undergoing glove integrity test)

Figure: 2 Rapid Transfer Port (RTP). Another means of transferring materials and tools

The gasket of the RTP systems has been identified as potential source of contamination in isolators since there may be a small contact surface around the gasket that may not be exposed to the decontaminating agent. A risk analysis should be done to evaluate potential contamination risk with the gaskets and their needs for maintenance program. Transfer of material into and out of the isolator is also a potential source of contamination.

Furthermore, Isolators may also be designed in combination with smaller enclosures associated with them to allow the continuous ingress of materials through the smaller Isolator into a main Isolator. Fully closed isolators are widely used for sterility testing aseptic process such as filling and crimping.

Classification Of Isolator Rooms

The surrounding room of an isolator should have limited access to staff (ensuring only the presence of authorised personnel), adequate space around the isolator and temperature/humidity under control for the effective utilization of decontamination technologies (e.g. Vapour Phase Hydrogen Peroxide systems).

Regulatory authorities require for background environment of aseptic production isolators to be classified at minimum in Zone (Grade) D (ISO 8 at rest)

There is, however, general consensus that sterility testing isolators need not be placed in a classified clean room, but it is important that such isolator surrounding room should be with restricted access for the personnel and should be monitored for temperature and humidity.

Isolator Decontamination

The purpose of bio-decontamination is to remove viable bioburden on exposed surfaces inside the isolator; a decontamination process should be performed using sporicidal chemical agents associated with decontamination equipment as gas/vapour phase decontamination system using hydrogen peroxide (e.g. VHP) or equivalent.

A de-contamination cycle is an automated machine cycle that is controlled and monitored during each stage of the cycle. Cycles can be divided into four stages:

-Dehumidification
-Conditioning
-Decontamination
-Aeration

Dehumidification: The dehumidification also known as pre-conditioning is designed to ensure that the isolator enclosure has is at a pre-defined humidity value (< 20 % RH) to ensure a proper concentration of decontaminating agent.

Conditioning: Depending on the complexity of the system, at a minimum, the isolator must have tightly controlled temperature range, positive pressure and air velocity control. During this initial stage, the isolator doors and ports must be closed and sealed. Any defects in the barrier system should result in an alarm and abort the cycle.

During conditioning, an automated leak test should be initiated to detect any breaks in the barrier system (e.g. defective gloves or seals). Heating of VHP delivery pipework also occurs.

The conditioning stage is when the decontaminating agent shall reach the minimum concentration required to achieve the desired microbial reduction.

Decontamination: At this stage the VHP is maintained in the isolator according to the dosing rate contained in the recipe or cycle settings.

The time and total amount of VHP must result in a kill in BIs placed within the isolator. Generally a 6 log reduction is required for a cycle to be deemed a success.

Aeration: During aeration stage the amount of residual decontaminating agent must fall to safe levels. (< 1ppm). This is done by blowing the hydrogen peroxide carrying air out of the barrier system using fresh air.

Recommended Critical Process Parameters	Typical Units
Amount of H2O2 during conditioning	(g)
Dosing rate (conditioning)	(g/min)
Time for conditioning	(mins)
amount of H2O2 during decontamination	(g)
Dosing rate decontamination	(g/min)
Time for decontamination	(mins)
Aeration time	(mins)

Decontamination Agents

Decontamination of isolators is achieved by the supply of gaseous sporicidal agents. These agents must capable of killing both bacterial endospores and fungal spores. The system typically turns liquid agents into a gaseous vapor.

The decontamination agent typically used in industry is hydrogen peroxide. Other agents include, formaldehyde, peracetic acid and chlorine dioxide. The rationale for selecting a particular agent should be based on technical data, sporicidal efficacy and the materials and products that come into contact with such agents.

Often the starting point when selecting an agent is the manufacturer's recommendations. Manufacturers of equipment trains are best positioned to understand interactions with seals and surfaces etc. In many cases, the equipment is designed with a particular type of decontamination agent in mind.

Another source of information is the datasheets provided by the agent manufacturers. Datasheets also give an insight into the suitability of a chemical based on its purity, concentration and safety.

The below factors should be considered with regards to bio-decontamination:

- ➢ Ensure as much surface area of components are exposed
- ➢ Minimise loads in order to limit the bioburden levels prior to the cycle starting.
- ➢ For filling and closing machines, design automation to ensure parts are moving during the cycle to facilitate exposure to the agent.
- ➢ Ensure all areas are dry and free of foreign objects and debris

Containment

Containment bioreactor systems designed for recombinant microorganisms require not only that a pure culture is maintained, but also that the culture be contained within the systems. Both GLSP and biosafety levels are detailed in this section.

GLSP (Good Large-Scale Practice) level of physical containment is recommended for large-scale research of production involving viable, non-pathogenic and non-toxigenic recombinant strains derived from host organisms that have an extended history or safe large scale use.

The GLSP level of physical containment is recommended for organisms such as those that have built-in environmental limitations that permit optimum growth in the large scale setting but limited survival without adverse consequences in the environment.
BL1-LS

(Biosafety Level 1 -Large Scale) level of physical containment is recommended for large-scale research or production of viable organisms containing recombinant DNA molecules that require BL1 containment at the laboratory scale.

BL2-LS

Level of physical containment is required for large-scale research or production of viable organisms containing recombinant DNA molecules that require BL2 containment at the laboratory scale.

BL3-LS

Level of physical containment is required for large-scale research or production of viable organisms containing recombinant DNA molecules that require BL3 containment at the laboratory scale.

No provisions are made at this time for large-scale research or production of viable organisms containing recombinant DNA molecules that require BL4 containment at the laboratory scale.

Process
and Cycle Development

Depyrogenation Tunnels

In many instances, process and cycle development not only serves to develop and verify machine parameters, but also serves as a learning and knowledge gathering experience.

Therefore, temperature mapping studies are valuable elements of any cycle develop with regards to Depyrogenation tunnels. In conducting temperature mapping, the temperature distribution profile of all sections of the tunnel can be learned. This can identify any potential "cold spots" or in general, provide confidence in the functionality of the equipment.

Other aspects of Depyrogenation tunnel cycle development is verifying the efficacy of the tunnel at its upper and low limits. In validation terms, this is referred to as Process- Operational Qualification (OQ-P), however, completing this during cycle development allows changes to be made more readily and free of the formal constraints of validation. Endotoxin test vials are used to show a reduction in levels as a result of Depyrogenation.

The following points should be addressed during cycle development:

> the minimum Depyrogenation hold times can be reached
> the minimum temperatures can be attained (270°C)
> exit temperatures are suitable for the product been manufactured or filled
> Temperatures within the tunnel are distributed evenly.
> Endotoxin challenge samples undergo a 3 log reduction
> Confirmation of a positive pressure cascade

For Cool Zone Sterilisation:

> A minimum temperature of 170°C can be reached and maintained for a minimum of 2 hours.
> Fh, lethality values meet acceptance criteria.
> Biological indicators show no growth after a defined incubation period.

Isolators

Critical to effective bio-decontamination cycles is the parameters and settings used in each stage of the cycle. As previously outlined, a bio-decontamination cycle has 3 stages (Conditioning, bio-decontamination and Aeration).

During cycle development, the process parameters that result in successful bio-decontamination need to be verified. Through the use of temperature mapping, geometric mapping, empirical data and process knowledge, worst case locations are selected and tested for VHP exposure (Chemical indicators) and bio-decontamination (Biological Indicators).

If a successful cycle is achieved, these parameters and settings can then be used during formal Validation studies (PQs), production and commercial manufacturing.

The amount of positions (locations) must also be defined. Small isolators such as transfer isolators may only have one or two dozen test locations. Larger isolators containing processing stations such as filling/capping can have many more test locations.

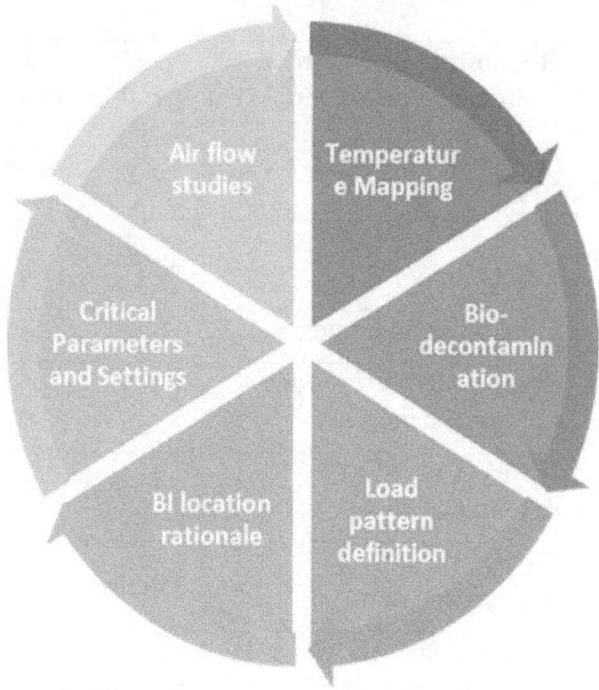

Figure 4: Key elements to Cycle Development and Verification.

Critical Parameters

Critical Process Parameters are parameters than can negatively affect the outcome of bio-decontamination if not adequately defined. Isolators typically have a high degree of automation that control and monitor air intake, dosing, temperature, air velocities and air pressures. The design and functionality of most isolators incorporate all of these conditions into the automated cycles or recipes.

Recommended Critical Process Parameters	Typical Units
Amount of H2O2 during conditioning	(g)
Dosing rate (conditioning)	(g/min)
Time for conditioning	(mins)
amount of H2O2 during decontamination	(g)
Dosing rate decontamination	(g/min)
Time for decontamination	(mins)
Aeration time	(mins)

Rationale for Locations

For a new isolator system, procedures and recipes must be put in place to support manufacturing operations. Prior to validation or the manufacturing of any saleable products, the isolator must be verified to ensure it operates as intended. A key part of this development phase is to define the locations or positions that will be used for the placement of indicators during testing (biological and chemical).

Locations can be selected under the below headings:

-Critical Positions
-Worst Case Positions
-Geometric Positions

Critical positions can include intervention points such as RTPs and glove surfaces. Any location where there is potential impact to product must be assessed and may merit the placement of BIs during bio-decontamination verification and eventually Performance Qualification.

Worst Case Positions (temperature) can be based on data generated from temperature mapping studies. Temperature mapping using Kaye Validators can identify where the temperature may vary or fall outside the range of typical temperatures.

Worst Case Positions (Indicator) from studies with chemical or biological indicators can also identify positions where air flow or VHP may not penetrate or reach effectively. These present higher risk areas that need to be assessed.

Geometric positions are positions that are selected to demonstrate the effectiveness across the area of the isolator. Air flow pattern analysis may also indicate high risk areas.

Temperature Mapping

Temperature mapping is used throughout Aseptic filling and manufacturing processes including Autoclaves, Ovens, Depyrogenation tunnels and isolators. Using equipment such as Kaye Validators and thermocouples, a clear and accurate temperature map can be developed.

For example, in order to temperature map a thermal oven, a select number or thermocouples or probes would be placed at pre-determined locations throughout the oven. The locations should represent the entire distribution of the oven and its boundaries.

So essentially, thermocouples would be placed on the walls, doors and shelves of the oven. A cycle can then be started and the temperature throughout the oven over the course of the cycle can be collected.

Load Pattern Definition

Isolator load patterns are made up tolls and materials that must be present within the isolator during manufacturing.

Examples are listed below. Multiples of the same item are common as the restricted nature of isolators requires positioning of such items in key positions to allow access through glove positions and so on.

Some common tools and materials:

- ➢ Slider tool
- ➢ Forceps (Multiple sizes)
- ➢ Scissors
- ➢ Settle Plates
- ➢ Contact Plates
- ➢ IPA Wipes
- ➢ Lint Free Dry Wipes
- ➢ Sterile Markers
- ➢ Sterile tape
- ➢ Ziploc Bags

Prior to the operation of an Isolator or Aseptic processing, a load pattern must be defined and documented. The load pattern is made up of tools and utensils that are needed during normal operation and production.

Common items include forceps, scissors, settle plates, zip-lock bags, sterile IPA wipes. All items must be wiped with IPA prior to placement within the isolator enclosure. Once defined and documented, the VHP cycle (bio-decontamination) must be validated for the isolator load pattern.

Points to note:

-Isolator should be clean and free of debris or rejects

-Materials and tools cleaned with IPA

-The Materials of construction (MOC) should be documented and compatible with VHP.

-The load pattern should be clearly defined in a Standard Operating Procedure or Work Instruction.

-Glove integrity is required in order to allow safe operation of the Isolator

-The gloves are pre tested to ensure the integrity of the gloves to allow safe and effective operation

Using VHP (Vapourised Hydrogen Peroxide) a bio-decontamination cycle is completed on the isolator to decontaminate the isolator and the contents of it (utensils, filling stations, capping stations etc.).

Once a successful VHP cycle is completed, the isolator and filling line are considered sterile and will remain sterile for the duration of the campaign. Autoclaved parts can be transferred aseptically in to the isolator during a campaign using rapid transfer ports (RTPS).The RTPs can be used to aseptically remove parts from the isolator also.

Smoke Studies on Depyrogenation Tunnels

Depyrogenation tunnels are logically positioned upstream to filling lines and container closure unit operations. As previously described, Depyrogenation tunnels process glass vials, ampules or cartridges to ensure they are sterile and endotoxin free.

A design requirement of tunnels is that they can create a Pressure differential inside and this can be maintained during normal operating conditions. The positive pressure differential is designed to ensure non sterile air does not enter the sterile cool zone of the Depyrogenation tunnel.

Smoke studies are performed to ensure proper airflow direction. The differential air pressure in the hot zone should always be positive relative to the infeed zone, cool down zone, and to the room or surrounding environment. Some studies also serve to demonstrate that there is no air ingress into the sterile cooling zone from outside the tunnel.

Equipment Validation

Overview

Validation is a legal and regulatory requirement for the manufacture of medicinal products. The area of Validation can be sub-divided into two elements. Equipment Qualification (EQ) and Process Validation. Equipment qualification ensure that equipment operates as intended and is installed in accordance with the manufacturers recommendation.

Process Validation involves the provision of documented evidence to confirm a particular process performs consistency and meets pre-determined specifications.

All equipment that can impact the quality of product is subject to Validation; hence equipment and systems used in aseptic manufacturing must undergo equipment and process validation.

Installation Qualification (IQ) protocols should cover verification that all utilities are installed correctly to the manufacturers recommendations. All sitting and mechanical connections should also be confirmed as adequate.

Other key tests and verifications includes:

- ➤ Documentation of Materials of Construction (MOC)
- ➤ Calibration of equipment based instrumentation
- ➤ Spare parts listing
- ➤ Preventative maintenance schedule creation
- ➤ Electrical installation verification
- ➤ Health and Safety assessment
- ➤ Ergonomic Assessment
- ➤ Documenting Software and Hardware
- ➤ Backup of software
- ➤ Backup of Recipes (Sterilization, Bio-decontamination etc.)

The system User Requirements Specification (URS) should provide the basis of testing and must be fulfilled during the course of Validation.

Materials of Construction (MOC)

The materials of construction and evidence of the same (certificates) forms part of Installation also. Materials must be fit for the intended purpose and compatible with products and manufacturing agents that come into contact with them.

For example, fermenters are made of materials that are suited to the use of steam sterilisation techniques and regular cleaning. Such materials can be classed as both non-reactive and non-absorptive surfaces. Most aseptic processing equipment that incorporates product contact surfaces are made of high grade stainless steel. Cheaper classifications of stainless steel can be used for jacketing and other non-product contact areas.

All interior product contact surfaces should be polished to a "mirror" finish. Welds also need to be finished in a similar manner. Electro polishing provides a better quality surface finish than mechanical polishing.

As with any chemical reaction, factors such as temperature, pH and oxygen concentration can impact the performance and yield. To ensure the optimum conditions are maintained, it is important to monitor and control such parameters and factors. By far the most common these days is automatic control of systems and equipment with automatic feedback and adjustment.

Operational Qualification (OQ) is the second component of Equipment Qualification. This is "Establishing by documented evidence that the equipment operates per specifications and over the required ranges and to required tolerances". Equipment is also tested to ensure alarms and controls operate as required and intended. Some typical checks included in an equipment-operational qualification are testing of alarms, control system testing, utility failures and functional and operational testing.

Suggested IQ/OQ Verifications / Tests

Standing Operating Procedures (SOPs)

SOPs are designed to provide formal documented instruction on how to execute tasks or operate equipment or machinery. While each company will require different headings, a work instruction or SOPs should cover set-up, system operation, cleaning and shutdown to name but a few.

Test Instrumentation Calibration

External test devices such as temperature probes, volt meters, lux meters and particle counters may be required to take measurements during an equipment qualification. Test instrumentation should have a suitable range, resolution and accuracy. Certificates of calibration should also be available, with calibration conforming to a recognized external standard. Information such as the serial number, model number and manufacturer should be recorded for reference. and traceability.

Equipment Based Instrumentation Calibration

All equipment based instruments must be calibrated as part of equipment validation or in advance of it. Instruments should have unique calibration ID.

Electrical Checks

Appropriate connections and earthing checks. Review of electrical drawings to ensure the physical status is as per drawings and specifications. Cables and electrical hazards should be also appropriately labelled.

Mechanical Checks

Ensure the systems are fixed, fastened and integrated mechanically. Safety guards and barriers should also be in place where required.

Pneumatic Checks

Verify the proper supply and integration of compressed air. Supply should be leak free, regulated with filters and water-traps fitted as required.

Documentation

Verification that design, operation and maintenance documentation has been received from the manufacturer and are stored appropriately.

Ergonomics

Controls and HMIs should be positioned to facilitate ease of use and should be identified clearly.

Health and Safety

Hazards are identified and guarded, pin points are identified. No trip hazards are evident and Emergency stops function.

Software

Equipment Installation software checks should record the names and version numbers of all software. HMI Software, PLC software, application software. Provision should be made for disaster recovery and backup.

Hardware

Computer hardware should be recorded to include the model, manufacturer, serial number and specification details.

Environmental

Any features detailed in the URS relating to environmental requirements need to be verified during IQ/OQ. For example, automatic shutdown after periods of inactivity. Heating and cooling systems should also be appropriately insulated.

Alarms

Automated processes such as sterilization tunnels, Autoclaves, Filling machines and Isolators typically have many alarms and controls. Alarms can be categorized as critical or non-critical to the process or product.

Depending on the vendor or manufacturer alarms can also be grouped according to the type of alarm (EHS, process, mechanical, pneumatic and so on) Alarms should be tested to ensure the right action by the machine is taken, the process comes to a safe stop, and that the alarm can be acknowledged and the alarm condition cleared.

Utility Failure

Also referred to as provoke testing, utility failure of compressed air, fume extraction, electrical supply and so is to ensure in the event of failure during commercial manufacturing, the equipment comes to a safe stop and can be brought back into use upon recovery of the utility.

Fixtures

Materials of construction must be suitable for the intended use. In aseptic processing, high grades of stainless steel (316L) are the preferred material of use.

Functional Tests

The individual functions of equipment must be verified during commissioning and qualification.

Depyrogenation Tunnels (Equipment Validation)

While many of the general Installation and Operational checks will be relevant to Depyrogenation tunnels, a number of key focus area's must be verified during Equipment IQ/OQ of Depyrogenation of Tunnels.

Temperature: IQ/OQ tests should verify the capability of the tunnel to reach and maintain set temperatures over the operational range. Overshooting or undershooting of set temperatures should not be observed.

The temperature should be consistent over the operational area of the tunnel in which product travels. Alarms must also be verified to ensure upper and lower alarm tolerances are responsive.

Pressure: For equipment qualification, pressure transducers and gauges must be calibrated and functionally verified. Maintaining positive pressure is a key element of Depyrogenation design.

Other Equipment IQ/OQ tests for Depyrogenation tunnels include:

➢ HEPA filter integrity
➢ Pressure differential control

Isolators (Equipment Validation)

In addition to the standard IQ/OQ equipment checks and verifications, the following list details Isolator specific verifications that should be considered:

➢ Air-flow pattern, empty, in normal operation and in bio-decontamination mode

- Air-flow pattern in empty and decontamination mode
- HEPA filter integrity
- Leak detection
- Leak test of isolator
- Maintenance of pressure cascades in multi-isolator system design.
- For isolator units configured with transfer isolators, maintenance of pressure differential that ensures the pressure in the workstation remains positive when passing items between the workstation and the transfer isolator.
- Confirmation of the injection rate with decontamination gas concentration
- Verification of the stage of recipe or cycle
- Temperature mapping
- Gas generation test

Facilities

Introduction

Facilities and utilities qualifications are typically pre-requisites to the validation of manufacturing equipment and systems. Much of the activity that deals with establishing a facility or building that is fit for purpose is managed under the broad heading of commissioning and qualification (C&Q).

Commissioning can be defined as the planned, documented, and managed engineering approach to the start-up and handover of facilities, systems, and equipment to the end-user. It must deliver a safe and functional environment that meets the pre-defined design and user requirements.

In strict terms, qualification is more concerned with the confirmation and documentation showing that equipment or systems are properly installed and functional. Qualification forms part of validation, but the individual qualification steps do not equal a validated process.

The establishment of a user requirements specification (URS) and detailed design specifications ensure that the building or facility will meet end users' needs and that it is fit for the intended purpose.

It also provides a level of protection to the contracting company responsible for the project or facility construction. Post-URS approval requires an approved Design Qualification (DQ).

This provides verification and a documented record that the proposed design is suitable for the intended purpose. Further verification including IQ/OQ/PQ should be applied as required based on the system impact and criticality of facilities/utilities.

Key Terms

Direct Impact: a system that can directly impact product quality.

Indirect Impact: where a system is not expected to directly impact the product quality but supports or is ancillary to a direct impact system.

No Impact: a system that does not directly impact product quality and does not support a direct impact system.

Alert limit: a value reached when the normal operating range of a critical parameter has been exceeded, indicating that corrective measures may need to be taken to prevent the action limit being reached.

At-rest: a condition where the installation is complete with equipment installed and operating in a manner agreed upon by the customer and supplier, but with no personnel present.

Cleanroom: an area (or room or zone) with defined environmental control of particulate and microbial contamination, constructed and used in such a way as to reduce the introduction, generation and retention of contaminants within the area.

Containment: a process or device to contain product, dust or contaminants in one zone, preventing it from escaping to another zone.

Contamination: the undesired introduction of impurities of a chemical or microbial nature, or of foreign matter, into or onto a starting material or intermediate, during production, sampling, packaging or repackaging, storage or transport.

In operation state: Condition where the installation is functioning in the defined operating mode with the specified number of personnel working in the manner agreed upon.

Trend analysis: Analysis of data in a given item of information over a period of time.

Point extraction: air extraction to remove dust with the extraction point located as close as possible to the source of the dust.

Pressure cascade: a process whereby air flows from one area, which is maintained at a higher pressure, to another area at a lower pressure.

Relative humidity: the ratio of the actual water vapour pressure of the air to the saturated water vapour pressure of the air at the same temperature expressed as a percentage. More simply put, it is the ratio of the mass of moisture in the air, relative to the mass at 100% moisture saturation, at a given temperature.

Turbulent flow: turbulent flow, or non-unidirectional airflow, is air distribution that is introduced into the controlled space and then mixes with room air by means of induction.

The level of qualification and validation testing required for any system should be based on a risk assessment, examining the criticality of the system and environment.

Risk assessments should consider the following points:

- ➢ Building design and construction features
- ➢ System boundaries and complexity
- ➢ Potential product impact

- ➢ Environmental controls and monitoring systems
- ➢ Potential impact to operator safety
- ➢ Type of qualification/validation (e.g. prospective, concurrent, or retrospective)

Controlled-not-classified (CNC) environments, utilities, and facility control systems also require adequate qualification/validation. Again, the impact on product quality should be determined in order to shape any validation. Routine monitoring test locations as well as alert and action levels should be determined in advance of any validation for environmental monitoring or utility systems.

The Displacement Concept (low pressure differential, high airflow)

This concept is commonly found in production processes where large amounts of dust are generated. Under this concept the air should be supplied to the corridor, flow through the doorway, and be extracted from the back of the cubicle.

Normally the cubicle door should be closed and the air should enter the cubicle through a door grille, although the concept can be applied to an opening without a door. The velocity should be high enough to prevent turbulence within the doorway resulting in dust escaping.

This displacement airflow should be calculated as the product of the door area and the velocity, which generally results in relatively large air quantities.

Note: This method of containment is not the preferred method, as the measurement and monitoring of airflow velocities in doorways is difficult.

Pressure Differential Concept (high pressure differential, low airflow)

The pressure differential concept may normally be used in zones where little or no dust is being generated. It may be used alone or in combination with other containment control such as a double door airlock.

The high pressure differential between the clean and less clean zones should be generated by leakage through the gaps of the closed doors to the cubicle. The pressure differential should be of sufficient magnitude to ensure containment and prevention of flow reversal, but should not be so high as to create turbulence problems.

In considering room pressure differentials, transient variations, such as machine extract systems, should be taken into consideration. A pressure differential of 15 Pa is often used for achieving containment between two adjacent zones, but pressure differentials of between 5 Pa and 20 Pa may be acceptable.

Where the design pressure differential is too low and tolerances are at opposite extremities, a flow reversal can take place. For example, where a control tolerance of $\pm$ 3 Pa is specified, the implications of rooms being operated at the upper and lower tolerances should be evaluated.

It is important to select pressures and tolerances such that a flow reversal is unlikely to occur. The pressure differential between adjacent rooms could be considered a critical parameter, depending on the outcome of risk analysis.

The limits for the pressure differential between adjacent areas should be such that there is no risk of overlap in the acceptable operating range, e.g. 5 Pa to 15 Pa in one room and 15 Pa to 30 Pa in an adjacent room, resulting in the failure of the pressure cascade, where the first room is at the maximum pressure limit and the second room is at its minimum pressure limit. Low pressure differentials may be acceptable when airlocks (pressure sinks or pressure bubbles) are used to segregate areas.

The pressure control and monitoring devices used should be calibrated and qualified. Compliance with specifications should be regularly verified and the results recorded. Pressure control devices should be linked to an alarm system set according to the levels determined by a risk analysis.

Manual control systems, where used, should be set up during commissioning, with set points marked, and should not change unless other system conditions change. Airlocks can be important components in setting up and maintaining pressure cascade systems and also to limit cross-contamination. Airlocks with different pressure cascade regimes include the cascade airlock, sink airlock and bubble airlock

GMP Zoning

Selecting a suitable classification for a room or manufacturing facility depends on several factors. Firstly, it can be said that sterile products require a more stringent set of criteria than non-sterile products. However, there is an extensive range of products and medical devices that are sterile but are used in different ways and consist of different materials and technology. Some sterile products are single use only and used for short term purposes and then disposed of.

Other sterile products are used subcutaneously for longer periods or even require implantation. Therefore, the design of a facility along with its HVAC specification must be appropriate to the product being manufactured. High risk products require greater control.

The goal of facilities and HVAC systems is to minimise contamination and the associated risks. Using a "sterile versus non-sterile" rule of thumb is not adequate when classifying a room or facility.

Standards including EN ISO 14644-1 and guidelines such as EU cGMP Guidelines EudraLex volume 4 Annex 1 (2008) should be consulted in order to fully understand the requirements of each ISO classification and grade of room.

ISO classifications do not specify room occupancy states but when a designation is applied, the occupancy state must be stated in the relevant documentation or procedure.

The most relevant European Guideline (Annex 1 of the EU cGMP Guideline) lists four classification grades and their associated particulate limits in the 'at rest' and 'in operation' conditions.

In general, for the sterile and non-sterile products, similar classes are applied, but in non-sterile production the producer could assign their classes, having similar particulate concentration, temperature, pressure etc. but lower air-change rate could be used.

Types of Contamination

> cross contamination (of a product/material with another product/material)
> non-microbial particulate contamination (non-viable particles)
> biological/microbiological contamination (viable particles/micro-organisms)
> Factors Influencing Contamination Cleanliness Levels in the Manufacturing Processes
> Process air cleanliness
> personnel hygiene and clothing
> work practices
> material design (material of construction, surface finishes, room finishes, equipment, open system/enclosed system utensils, etc.)
> material cleanliness

Compliance Tests for GMP Zones

Test	Requirements
Particle count test	Test covers verification of cleanliness. Dust particle counts to be carried out and result printed. The number of readings and positions of tests should be defined in accordance with ISO 14644-1 Annex B5
Air pressure difference	This test is used to verify non cross-contamination. Log of pressure differential readings to be produced or critical plants should be logged daily, preferably continuously. A 15 Pa pressure differential between different zones is recommended. Refer to ISO 14644-3 Annex B5
Airflow volume	To verify air change rates. Airflow readings for supply air and return air grilles to be measured and air change rates to be

	calculated. Refer to ISO 14644-3 Annex B13
Airflow velocity	To verify unidirectional flow or containment conditions. Air velocities for containment systems and unidirectional flow protection systems to be measured. Refer to ISO 14644-3 Annex B4
Filter leakage tests	To verify filter integrity. Filter penetration tests to be carried out by a competent person to demonstrate filter media, filter seal and filter frame integrity. Only required on HEPA filters. Refer to ISO 14644-3 Annex B6
Containment leakage	To verify absence of cross-contamination. Demonstrate that contaminant is maintained within a room by means of: • airflow direction smoke tests • room air pressures. Refer to ISO 14644-3 Annex B4
Recovery	To verify clean-up time.

	Test to establish time that a cleanroom takes to recover from a contaminated condition to the specified cleanroom condition. Should not take more than 15 minutes. Refer to ISO 14644-3 Annex B13
Airflow visualisation	To verify required airflow patterns. Tests to demonstrate air flows: • from clean to dirty areas • do not cause cross-contamination • uniformly from unidirectional airflow units Demonstrated by actual or video-taped smoke tests. Refer to ISO 14644-3 Annex B7

HEPA Filters

HEPA filters are composed of a mat of randomly arranged fibres. The fibres are typically composed of fiberglass and possess diameters between 0.5 and 2.0 micrometers. Key factors affecting its functions are fibre diameter, filter thickness, and face velocity. The air space between HEPA filter fibres is typically much greater than 0.3 μm.

The common assumption that a HEPA filter acts like a sieve where particles smaller than the largest opening can pass through is incorrect and impractical.

Unlike membrane filters at this pore size, where particles as wide as the largest opening or distance between fibres can not pass in between them at all, HEPA filters are designed to target much smaller pollutants and particles.

These particles are trapped (they stick to a fibre) through a combination of the following three mechanisms:

Interception

Where particles following a line of flow in the air stream come within one radius of a fibre and adhere to it.

Impaction

This is where larger particles are unable to avoid fibres by following the curving contours of the air stream and are forced to embed in one of them directly; this effect increases with diminishing fibre separation and higher air flow velocity.

Diffusion

Diffusion is an enhancing mechanism that is a result of the collision with gas molecules by the smallest particles, especially those below 0.1 μm in diameter, which are thereby impeded and delayed in their path through the filter; this behavior is similar to Brownian motion and raises the probability that a particle will be stopped by either of the two mechanisms above; this mechanism becomes dominant at lower air flow velocities.

Diffusion predominates below the 0.1 μm diameter particle size. Impaction and interception predominate above 0.4 μm. In between, near the most penetrating particle size (MPPS) 0.21 μm, both diffusion and interception are comparatively inefficient. Because this is the weakest point in the filter's performance, the HEPA specifications use the retention of particles near this size (0.3 μm) to classify the filter.

However it is possible for particles smaller than the MPPS to not have filtering efficiency greater than that of the MPPS. This is due to the fact that these particles can act as nucleation sites for mostly condensation and form particles near the MPPS.

Lastly, it is important to note that HEPA filters are designed to arrest very fine particles effectively, but they do not filter out gasses and odor molecules.

The specification usually used in the European Union is the European Norm EN 1822:2009. It defines several classes of HEPA filters by their retention at the given most penetrating particle size (MPPS):

Today, a HEPA filter rating is applicable to any highly efficient air filter that can attain the same filter efficiency performance standards as a minimum and is equivalent to the more recent NIOSH N100 rating for respirator filters.

The United States Department of Energy (DOE) has specific requirements for HEPA filters in DOE regulated applications. In addition, companies have begun using a marketing term known as "True HEPA" to give consumers assurance that their air filters are indeed certified to meet the HEPA standard.

The specification usually used in the European Union is the European Norm EN 1822:2009. It defines several classes of HEPA filters by their retention at the given most penetrating particle size (MPPS):

HEPA class	retention (total)	retention (local)
E10	> 85% ---	
E11	> 95% ---	
E12	> 99.5%	---
H13	> 99.95%	> 99.75%
H14	> 99.995%	> 99.975%
U15	> 99.9995%	> 99.9975%
U16	> 99.99995%	> 99.99975%
U17	> 99.999995%	> 99.9999%

Today, a HEPA filter rating is applicable to any highly efficient air filter that can attain the same filter efficiency performance standards as a minimum and is equivalent to the more recent NIOSH N100 rating for respirator filters.

The United States Department of Energy (DOE) has specific requirements for HEPA filters in DOE regulated applications. In addition, companies have begun using a marketing term known as "True HEPA" to give consumers assurance that their air filters are indeed certified to meet the HEPA standard.

Performance Validation

Depyrogenation -Performance Qualification (PQ)

Performance Qualification examines the effectiveness and reproducibility of the Depyrogenation cycles in respect of a particular load. Thermocouples are used to verify the set temperatures are reached and maintained throughout the cycle and at various points in the tunnel to demonstrate consistent heat distribution. Endotoxin challenge vials or ampules are also used to demonstrate reduction in endotoxin levels to within acceptable levels and hence ensuring products are safe for patient use.

Traditionally, the theme of consistency in Performance Qualifications has been demonstrated by completed 3 distinct batches or runs as part of PQ. Although Risk based approaches to validation (e.g. ASTM E2500 etc.) are increasingly used across the life science industry. Completing a minimum of 3 batches or runs for initial Performance Qualification is still the expected requirement.

The size of the glass components (vials, ampule's or cartridges) must also be considering in the design of Performance Qualifications. The efficacy of the tunnel or sterilising effect is ultimately determined by the size, shape and mass of the components processed. A family or bracketing approach may be utilised to reduce the amount of runs or cycles required.

For example, if a manufacturing process utilising 4 different vial sizes across a product – 5ml, 10ml, 15ml and 20ml. Based on technical rationale and some level of evidence, the 5ml and 20ml vials could be considered "worst case". The 5ml been the smallest may exhibit the smallest nick sizes and internal geometry. The 20ml would in this case be the largest vial, with the largest surface area and mass – another worst case configuration.

Another consideration is the position and quantity of thermocouples. For PQ, a rationale based on sound science must support the locations and quantity of thermocouples used during the validation.

This should be based on historic data if available, or in the case of new equipment, data generated during SAT testing and/or Engineering development studies.

The thermocouple placement should also be carefully considered. If a particular location is to be assessed, the thermocouple should be secured with Kapton tape. Movement of the probe during a test cycle may result in the data been inconclusive or deemed non-representative.

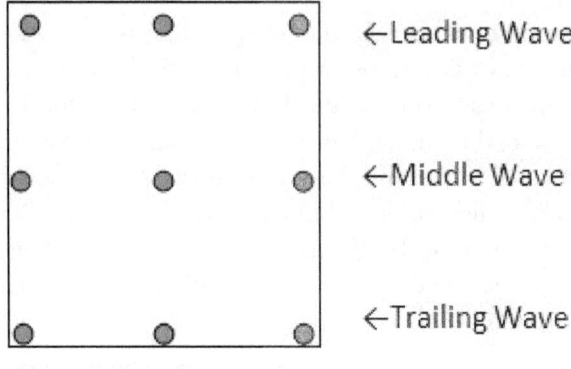

= Thermocouple

Figure 9: Simple representation of a Depyrogenation tunnel (rectangle). Tunnels are filled with components which progress along a belt. Therefore, each batch shall have a leading wave, middle wave and trailing wave. Place thermocouples at the above locations should verify the temperatures at each location as the product moves through the Depyrogenation tunnel. Endotoxin challenge samples may also be placed in similar locations.

Thermocouples – Points to consider

- ➢ The point of contact with the component (e.g. bottom of vial, middle of vial etc.)
- ➢ Size of component if a bracketing approach is chosen
- ➢ Are there cold spots within the tunnel
- ➢ Rating of thermocouples (Depyrogenation temperatures can well exceed 290°C).
- ➢ Length of thermocouples should allow for the travel through the length of the Depyrogenation tunnel.
- ➢ Number of locations probed (large Depyrogenation tunnels require increased data points)

Cool Zone- Performance Qualification

Depyrogenation Zone	Sterile Cool Zone

Direction of Product travel through Depyrogenation Tunnels

→

As illustrated above Performance Qualification of Depyrogenation tunnels can be divided into two elements (1) Depyrogenation and (2) Cool Zone Sterilisation. After glass vials or ampules are processed through the Depyrogenation zone, they then enter the "Cool Zone" where temperatures are typically lower.

However, prior to the transfer of product into the cool zone, a sterilisation cycle is typically performance. This consists of setting the Cool Zone temperature to known temperature for a defined period of time. For example a Cool Zone sterilisation cycle may require at set temperature of 230°C for 2 hours.

For initial Performance Qualifications 3 consecutive runs are typically completed for newly installed equipment. Cool Zone PQs are completed using temperature mapping equipment such as Kaye Validators. Specific locations are probed with temperature probes and a Cool Zone sterilisation Cycle is run.

The positions of thermocouples should be based on technical rationale developed in earlier process and cycle development. The positions must be representative of the size and shape of the tunnel. For example positions that may merit the placement of a heat penetration thermocouples may include:

-Tunnel Belt
-Interior Panels /Walls
- Doors (Seperation Depyrogenation zone from Cool Zone.)
-Extremities of the Cool Zone

BIs can also be placed in positions identified as "cold spots". If the Sterilisation cycle is effective in rendering biological indicators located at the coolest positions sterile, then it can be concluded that positions showing a higher temperature are also found to be sterile.

Recommended Depyrogenation Performance Qualification Tests

Parameter/Description	Recommended Acceptance Criteria
Depyrogenation Hold-time: defined as the total time in which the tunnel reaches and maintains a set temperature during a cycle, where the temperature of vials is held at a minimum of 270°C	≥102 seconds at 270°C
Endotoxin Challenge: a reduction in levels of Endotoxin (of Endotoxin spiked samples)	≥ 3 log reduction
Exit temperature: the temperature of vials/ampules upon exiting the coolzone into the infeed of the isolator.	≤25°C or as required based on the product. Some proteins and treatments are sensitive to moderate temperatures

Recommended Cool Zone Sterilisation Performance Qualification Tests

Parameter/Description	Recommended Acceptance Criteria
Dry Heat Hold-time: the total time in which all locations in the tunnel are ≥ 170°C.	≥120mins at 170°C
Fh Calculation: a measure of sterilisation time	≥ 32 mins

F value Definition

A term used to describe the delivered lethality calculated based on the physical parameters of the cycle (e.g., F0, FH). The F-value is the integration of the lethal rate (L) over time: The lethal rate is calculated for a reference temperature (T ref) and z-value using the equation L=10(TT ref)/Z.

Lethality Definition

The capability of the sterilisation process to destroy micro-organisms, i.e. achieve a 106 reduction in population of Bacillus Stearothermophilus.
☐

Isolators

With reference to PIC/S guidance (PI 014-3, Isolators used for Aseptic Processing and Sterility Testing) placing three or more BIs at defined worst case locations is recommended for performance qualification. Depending on the size of the isolator, BI locations may range from 20 to 90 different positions or more.

If there are any positive growths the proportion of positive to negative can be used to estimate the number of survivors and thus the calculate log reductions. Given this information any variation in the process is estimated and the significance of it can be evaluated. If there is only one BI in each position, and only growth/no growth is established, then the number of any survivors is unknown and the size of the possible variation in the process cannot be estimated.

The BI challenges demonstrate that the VHP cycle parameters to be qualified are sufficient to ensure a six log reduction in a known microbiological system.

> ➢ BI Challenge is achieved by placing Biological Indicators within the barrier system.
> ➢ Chemical Indicators are included. These change colour when exposed to the decontamination agent.
> ➢ Temperature readings are captured with Data loggers
> ➢ Relative Humidity readings are captured with Data loggers.

➢ Reduced Process Parameters are controlled via a recipe.

Reduced Parameters

As an output of any Engineering development (aka cycle development) a production recipe or cycle should be documented. One would have proven the efficacy of the recipe or cycle during the testing or runs completed. However, regulations require manufacturing processes to be validated. Once the pre-requisites to validation are completed (equipment IQ/OQ and Engineering/cycle development) Performance of the Isolator barrier system is required.

Execution Performance Qualification at worst case parameters is achieved by reduced one or more critical parameters to less than normal production parameters. For example reducing the amount of VHP dosed during the bio-decontamination stage of a cycle would result in a greater challenge. (Less VHP may lead to a positive BI result or no change in Chemical indicators).

If a production recipe has a dosing time of 20 minutes, reducing the dosing time to 15 minutes during validation runs would constitute a "worst case".
Isolator Load Verification
To ensure that the load pattern qualified under performance qualification is accurate and according to established procedures (from engineering and cycle development) load verification should be completed.

This simply means the items that make up the load are placed in there correct positions within the isolator. Visual confirmation that the right tools and materials are in the right positions, at the right quantities. Load components at the exact locations as detailed in SOPs or the relevant validation protocol (e.g. Performance Qualification Protocol)

Isolator VHP Biological Indicator Challenge

VHP BI challenges are used to demonstrate that the VHP cycle parameters are capable of achieving a six log reduction from a known microbial population.

In order to prove consistency, BIs (Biological Indicators) are often placed in triplicate at predefined positions with the isolator. Once again, these positions should be based on scientific rationale and backed by documented evidence from either engineering runs or process development studies

After a bio-decontamination cycle has finished, test BIs must be removed from the isolator. As the BIs should now be rendered "sterilized", care should be taken in order to prevent contamination through poor handling practices.

- ➢ Wear sterile gloves and sanitise with 70% IPA
- ➢ Allow IPA to dry prior to removing BIs from BI location.
- ➢ Use aseptic technique transfer BIs into sterile centrifugal tubes.
- ➢ Sanitise hands with 70% IPA after BI is removed and prior to handling the next.

Chemical Indicator Challenge

The chemical indicator study will be performed for each run to verify uniform exposure of VHP through the Filling and Closing Machine Isolator.

Chemical indicators are placed throughout the Filling and Closing Machine Isolator with each group of triplicate BIs. The VHP cycle will be run and once the run is complete the CIs will be removed and inspected. Chemical indicators yield qualitative information by changing colour.

The indicators generate qualitative information enabling verification that the decontamination agent has reached all the points to be decontaminated.
All the chemical indicators shall display colour change at the end of the decontamination phase to indicate exposure to VHP.

Temperature and Humidity Study
The purpose of temperature and humidity mapping is to record the temperatures and humidity within an isolator during the conditioning and bio-decontamination phases of the cycle. Both temperature and humidity can influence the effectiveness of a cycle. A well designed Isolator will ensure a relatively stable temperature and humidity in particular at the pre-conditioning stage. The outer environment in which the Isolator sits can also impact both temperature and humidity.

During process development, Kaye Validators and thermocouples may have been used to record temperatures. However, during Validation of Isolators, Data Loggers are more suitable. Data loggers can be placed in pre-determined positions to capture both temperature and humidity.

Pre-requisites to execution of a Bio decontamination Run

> ➤ Record the lot number and item number of the VHP sanitant
> ➤ Record the expiry date of the VHP
> ➤ Confirm Isolator has been sanitized using IPA 70/30
> ➤ Record the lot number and item number of the IPA 70/30
> ➤ No debris

H2O2 Monitoring

Vapourised Hydrogen peroxide (H2O2) used during bio-decontamination has the potential to cause harm to personnel and operators working in the vicinity of isolators. Ambient H202 levels are monitored using extremely responsive sensors that pick up minute traces of VHP in the atmosphere (to within 2-10 part per million (ppm)).

If a VHP sensor picks up levels of VHP within its working envelope, it typically triggers an alarm. The isolator system should also come to safe stop. For example, if the cycle is in the bio-decontamination stage where VHP is been delivered to the isolator, the system should immediately stop this process and safely abort the cycle.

Internal H_2O_2 sensors also allow ongoing monitoring of concentrations during a cycle. Alert levels and action limits are typically defined in a cycle recipe when is used to define the duration and parameters requirements.

Critical Parameters Monitoring

Critical parameters such as humidity, pressure, temperature and air velocity should be documented in an approved and controlled SOP. It is also important to verify parameters prior to initiating a cycle. Batch reporting by automated systems also facilitate the review of critical parameters. In particular, temperature and relative humidity are important parameters to regularly monitor and trend both in the isolator and the surrounding room during decontamination of isolators.

Thermal Validation Systems

Performance Qualification for both Depyrogenation tunnels and isolators require the use of thermal validation systems. These calibrated systems capture temperature data using heat penetrating thermocouples. In advance of conducting a temperature mapping study, the equipment needs to be set up correctly. Most modern systems allow you to define a study in which the temperature range, midpoint and other calibration information is selected. Calibration verification is also completed post study execution to ensure the accuracy of the data.

Re-Qualification

Depyrogenation: Performance Validation for Depyrogenation tunnels are normally divided into two separate studies of validation protocols- (1) Depyrogenation Performance Qualification and (2) Cool Zone Sterilisation Performance Qualification. Annual re-qualification by completed one run is acceptable to ensure the equipment is still effective in removing neutralizing pyrogens and provide sterilization.

Isolators: Re-qualification activities are typically limited to conducting one PQ decontamination run on an annual basis. Requalification should also ensure the load pattern and other critical parameters are in control. If amendments are needed such as a change to a loading pattern or quantity of items, then this may require further validation. The impact of the change should be assessed and documented. The purpose of the re-qualification is to confirm the performance of the decontamination cycle is still effective since the initial validation runs.

Culture Media

Culture media is used to fill vials or ampoules in order to evaluate the viable contamination risks from the environment including the personnel. Culture media is a sterile media however when processed through a filling and closing machine will support the growth of microbes if there are issues with the line sterility.

Media fills are studies completed on fill lines in order to demonstrate that Sterility can be achieved and maintained during the course of normal manufacturing. Media fills typically are done in advance of Process Validations.

☐

A

Accelerated Aging

When the deterioration of a device or product component from natural aging is accelerated and simulated in the laboratory.

Accuracy

Accuracy or trueness. An expression of the closeness of agreement between the value that is accepted, either as a conventional true value or an accepted reference value and the value obtained. A system with low bias implies good accuracy and vice versa.

Adverse Event

A situation or condition that occurs when a data point, result, or process etc. is outside the expected or predetermined limits or ranges.

Air Exchange Rate Per Hour (ACPH)

The rate of air exchange expressed as number of air changes per hour and calculated by dividing the volume of air delivered in the unit of time by the volume of space.

Active Pharmaceutical Ingredient

Any substance or mixture of substances intended to be used in manufacturing a drug (medicinal) product and that, when used in the production of a drug, becomes an active ingredient of the drug product. Such substances are intended to furnish pharmacological activity or other direct effect in the diagnosis, cure, mitigation, treatment, or prevention of disease, or to affect the structure and function of the body. (ICH Q7A, Annex 18, Part II)

ANSI

American National Standards Institute

Antimicrobial Resistance

Antimicrobial resistance corresponds to the emergence and spread of microbes that are resistant to cheap and effective first-choice, or "first-line" antimicrobial drugs.

Application

A term most often used in relation to Software validation and computerized systems. It is any software installed on a defined platform providing specific functionality.

Approve

"Approve" the device after reviewing a premarket approval (PMA) application that has been submitted to FDA.

AVL (Approved vendor list)

A list of all the vendors or suppliers approved by a company as sources from which to purchase materials.

Artwork

Electronic files or printouts containing the representation of a packaging item, graphical elements, and regulatory text. Approved artworks are used by suppliers for printing.

Aseptic (conditions)

Conditions in the working environment under which the potential for microbial and/or viral contamination is minimized.

ASTM

American society for testing and materials.

ATEX

ATEX, an acronym of the French Atmospheriques Explosives. This European Directive amends and adds safety requirements for hazardous areas in the relevant national legislation in the member states of the European Union, bringing in a common standard.

Where equipment is to be used in potentially explosive atmospheres containing gas or combustible dust, it must comply with the ATEX directive.

Audit Trail

The audit trail is a control mechanism of a system that allows all data entered or modified to be traced back to the original data. A reliable and secure audit trail is particularly important in conjunction with the creation, change or deletion of GMP relevant electronic records.

Acceptable Quality Level (AQL)

The AQL of a sampling plan is the Process Performance Level routinely accepted by the sampling plan.

B

Basis of Design

A design document that demonstrates a thorough understanding of the project and its intended output. Typically contains preliminary drawings and system descriptions etc. Together with the URS and the Detailed Design, it provides overall evidence that the design addresses the requirements of the equipment, system or facility.

Biocompatibility

A measure of how a biomaterial interacts in the body with the surrounding cells, tissues and other factors.

Bioburden

The level and type of micro-organisms that can be present in raw materials, API starting materials, intermediates or APIs. Bioburden should not be considered contamination unless the levels have been exceeded or defined objectionable organisms have been detected.

Biological Indicators

Test system containing viable microorganisms providing a defined resistance to a specified sterilization process, e.g. Vaporised hydrogen peroxide.

Biomaterial

Any matter, surface, or construct that interacts with biological systems. Biomaterials can be derived from nature or synthetic (manufactured).

The active substance of a biosimilar medicine is comparable to a biological reference medicine. Biosimilar and biological reference medicines are used at the same dose to treat the same disease. The name, appearance and packaging of a biosimilar medicine differs to that of a biological reference medicine.

Bracketing

A Bracketing (aka family or matrix) approach can be used where similar products are produced using the same equipment and processes. A particular product size or product configuration may be selected to represent the worst-case product. Therefore, by qualifying the worst case, all of the other products within the family are considered validated.

Body orifice

Any natural opening in the body, as well as the external surface of the eyeball, or any permanent artificial opening, such as a stoma or permanent tracheotomy.

Borderline Classifications

In certain circumstances, it may not be clear if a product falls under the medical device legislation or whether to classify a device as a medicine, cosmetic, biocide and so on. The decision will largely depend on the particular intended use of the product, as assigned by the manufacturer, and on the demonstrated mode of action. The manufacturer's claims must be substantiated by relevant data.

Bulk Product

Any pharmaceutical form (liquid, powder, suspension) that is to be filled into either another container or its final container at the next process step; or is already filled into its final container to be labelled and packaged at the next process step.

BOM

Bill of Materials.

BSI

British Standards Institute.

C

CAD, Computer Aided Drawing

A system used to create physical designs, usually three-dimensional. Some examples of CAD software are SolidWorks, Pro/ENGINEER and AutoCAD.

Calibration

The a requirement that demonstrates a particular instrument or device produces results within specified limits by comparison with those produced by a reference or traceable standard over an appropriate range of measurements.

Campaign (Process)

A production strategy where consecutive batches of an API, a finished product, or intermediates are processed before the production line/system is cleaned.

Capability (Process Capability)

Process Capability is a measure of how capable the process is of producing product meeting specified requirements. It is a measure of the actual variation in that product characteristic compared to the product specifications.

Indices are used to represent the Process Capability such as Pp, Cp and Ppk, Cpk, depending on how the data is collected e.g. multiple batches over time.

CAPA

A Corrective and preventive action. A systematic approach that includes actions needed to correct, prevent recurrence and eliminate the cause of potential nonconforming product and other quality problems (preventive action) (21CFR 820.100)

Change Control

A formal system by which qualified representatives of appropriate disciplines review proposed or actual changes that may impact the validated status.

Change Notification (Agreement)

A signed declaration that states that the Supplier agrees to notify the customer of changes in its product or process in order to allow the customer determine whether the changes can affect the quality of finished goods or Quality System.

Change Management

An overarching approach to change control that is used during the preliminary planning and design stage of a project.

Cleaning

The process of removing potential contaminants from process equipment and maintaining the condition of equipment such that the equipment can be safely used for subsequent product manufacture.

Cleaning Validation

Documented evidence that provides a high degree of assurance that a specific cleaning process will consistently produce a result meeting predetermined requirements for cleanliness.

Cleaning Verification

Confirmation by examination and provision of objective evidence that specific requirements have been fulfilled.

Cocurrent (flow)

This is when the fluids are applied in the same direction.
Cocurrent flow is less effective as less heat can be transferred, therefore it is less commonly used.

Code of Federal Regulations (CFR)

Regulations issued by U.S. government agencies. The individual titles making up the regulations are numbered the same way as the federal laws on the same topic.

Competent Authority

A competent authority is the legally designated authority mandated to monitor compliance with directives and legal requirements within the industry. The competent authority has the power to grant and revoke licenses.

Compendial Organisations

Organizations certifying material standards that meet compendial requirements and acceptance criteria. (e.g. USP).

Commissioning

An engineering activity that includes all aspects of bringing a system, piece of equipment or process is installed and ready for use. Commissioning involves both requirements of Installation Qualification (IQ) and Operational Qualification (OQ).

Computer System

A group of hardware components and associated software, designed and assembled to perform a specific function or group of functions.
[EU GMP Guide, Part II, ICH Q7]
Computerised System

A system including the input of data, electronic processing and the output of information to be used either for reporting or automatic control. [EU GMP Guide, Glossary]

Computer System Validation
A process that confirms by examination and provision of objective evidence that the computer system conforms to user needs and intended uses. System validation is a process for achieving and maintaining compliance with GxP regulations and fitness for intended use by adoption of life cycle activities, deliverables, and controls.

Concurrent Validation
Concurrent Validation occurs when activities are executed at the same time as one another or concurrent to a product launch.

Confidence Level
Confidence Level is expressed as a percentage and represents the probability that the conclusion of the test is correct. A 95% confidence level means you can be 95% certain that the conclusion is correct.

Conflict Of Interest
A conflict of interest is a situation in which a public official's decisions are influenced by the official's personal interests.

Continual Improvement, CI

Ongoing activities to evaluate and positively change products, processes, and the quality system to increase effectiveness

Consent Decree

A consent decree is a binding order issued by a judge that stipulates the voluntary agreement by the participants in a case of litigation. Decrees are sometimes issued after one party voluntarily agrees to cease a particular action without admitting to any illegality of the action to date.

Colony Forming Unit

One or more microorganisms that produce a visible, discrete growth on an agar-based microbiological medium.

Controlled Substances

Products that are categorized due to their potential for abuse, medical use and requirement for medical supervision.

Controlled classified areas

An environment supplied with HEPA filtered air where materials, equipment, and personnel are regulated to control viable and non-viable particulates to an acceptably low level. Such areas are classified according to the maximum level of airborne particulate allowed.

CNC (Controlled Not Classified)

While these are not ISO recognized room classes, they are generally used to describe non-GMP areas with a level of control in effect.

Clear (FDA)

"clear" the device after reviewing a premarket notification, otherwise known as a 510(k) (named for a section in the Food, Drug, and Cosmetic Act), that has been filed with FDA, or

Cleanroom

An area (or room or zone) with defined environmental control of particulate and microbial contamination, constructed and used in such a way as to reduce the introduction, generation and retention of contaminants within the area.

Containment

A process or device to contain product, dust or contaminants in one zone, preventing it from escaping to another zone.

Contamination

The undesired introduction of impurities of a chemical or microbial nature, or of foreign matter, into or onto a starting material or intermediate, during production, sampling, packaging or repackaging, storage or transport.

Continued Process Verification

Once the initial validation is completed it is important that the system or process remains within the validated state. This is done by monitoring the performance and output of the system or equipment. Furthermore, any changes to this system or equipment must be assessed and documented in order to assure the product is safe and meets acceptance criteria.

Critical Aspects

Critical aspects of manufacturing systems include the functions, features, abilities, and performance or characteristics required for the manufacturing process and systems to ensure consistent product quality and patient safety. They should be identified and documented based on scientific product and process understanding.

Critical Quality Attribute, CQA (Critical-to-Quality)

A property or characteristic with specific nominal value and appropriate limit and range providing a particular quality attribute. A CQA typically is classed as a high risk requirement, where the safety or efficacy of the product depends on the CQA been within the specified limits.

CCC (Mark)

The "China Compulsory Certificate"mark, commonly known as CCC Mark, is a safety mark for many products sold on the Chinese market. As of 2013, medical devices do not require this certification.

CDC
Center for Disease Control & Prevention (USA)

CDRH

Center for Devices and Radiological Health (USA)

CE Marking

The CE Marking is a mandatory conformance mark on many products (including medical devices) placed on the single market in the European Economic Area. The CE marking certifies that a product has met EU consumer safety, health or environmental equirements. By affixing the CE marking to a product, the manufacturer declares that it meets EU safety, health and environmental requirements.

CEN

Communité Européenne des Normes (European Committee for Standardization).

Clinical Trial

Clinical Trials are conducted to allow safety and efficacy data to be collected for health interventions (e.g., drugs, diagnostics, devices, therapy protocols). These trials can take place only after satisfactory information has been gathered on the quality of the non-clinical safety, and Health Authority/Ethics Committee approval is granted in the country where the trial is taking place.

Clinical Trial Sponsor

The Clinical Trial Sponsor is responsible for the safety of subjects in a clinical trial and informs local site investigators of the true historical safety record of the drug, device or other medical treatment to be tested, and of any potential interactions of the study treatment(s) with already approved medical treatments.

Cleaning

Removal of contamination or soils from an item or surface to the extent necessary for its further processing and its intended subsequent use.

CMDCAS

Canadian Medical Devices Conformity Assessment System.

CMDR
Canadian Medical Device Regulation.

Conformity

Fulfilment of a requirement or meeting a requirement.

Conformity Assessment Body (CAB)
A body, other than a Regulatory (competent) Authority, engaged in determining whether the relevant requirements in technical regulations or standards are fulfilled.
CRO
A "Contract Research Organization", also commonly known as a "Clinical Research Organization", is a service organization that provides support to the pharmaceutical and biotechnology industries. CROs offer clients a wide range of "outsourced" pharmaceutical research services to aid in the drug and medical device research and development process.

D

Data Integrity

Is the degree to which data is reliable and without error. Data must be accurate, attributable, contemporaneous, original, legible and available. A breach of data integrity occurs when any person manipulates or distorts data and submits the results of that data as valid.

Dead Leg

A dead leg in the world of piping terminology refers to an area of piping where there is insufficient flow or a tendency for water build-up or stagnation.
The formal definition of a dead-leg states that
Pipelines for the transmission of purified water for manufacturing or final rinse should not have an unused portion greater in length than 6 diameters (6D rule) of the unused portion of pipe measured from the axis of the pipe in use.

Debugging

The process of locating, analysing, and correcting suspected faults or machine issues.

Design controls

Design controls are a collection of practices and procedures that are incorporated into the design and development process for a product such as a medical device. It provides a structure and clear path from user needs assessment to product delivery through a step-by-step process. Design controls ensure proper assessment of the design is completed during the design and development phase. Design controls are a requirement of quality systems such as 21 CFR Part 820 (medical devices), and for certain classes of devices and per ISO 13485 - Quality Management Systems.

Decommissioning

When a system is taken out of production service and stored in an adequate environment for potential future use.

Depyrogenation

A thermal process used to destroy or remove pyrogens (endotoxins). Typically primary packaging components such as glass vials are subject to Depyrogenation.

Detection Limit

The lowest amount of analyte in a sample that can be detected but not necessarily quantitated as an exact value for an individual analytical procedure. (Ref: ICH Q2)

Design History File
The DHF is a repository for all of the documentation generated as a result of the design control process. The DHF serves as a complete record of the design.

Design Validation
Establishing by objective evidence that device or product specifications conform to user needs and intended use(s) defined in design documentation.

Debarment

The FDA has the authority to "disqualify," or remove, researchers from conducting clinical testing of new drugs and devices when the agency determines that the researcher has repeatedly or deliberately not followed the rules intended to protect study subjects and ensure data integrity. Further, the FDA can disqualify a clinical investigator who has repeatedly or deliberately submitted false information to the agency or study sponsor in a required report.

Under its statutory debarment authority, the agency may also ban, or "debar" from the drug industry individuals and companies convicted of certain felonies or misdemeanours related to drug products. Once individuals have been subjected to "debarment," they may no longer work for anyone with an approved or pending drug product application at FDA. Debarred companies may no longer submit abbreviated drug applications.

Design qualification (DQ)

The documented verification that the proposed design of the is suitable for the intended purpose. DQs are typical deliverables for facilities, systems and equipment and or processes.

Design Space

The multidimensional combination and interaction of input variables, e.g. material attributes, and process parameters that have been demonstrated to provide assurance of quality. Working within the design space is not considered as a change.

Directives

Directives are legal requirements. These must be met by manufacturers. Standard such as ISO 13485 help companies meet the requirements of directives, such as "Guidelines Relating to the Application of the Council Directive 93/42/EEC on Medical Devices."

Direct impact (system)

A system that is expected to have a direct impact on product quality. These systems are designed and commissioned in line with Good Engineering Practice (GEP) and, in addition, are subject to Qualification and Validation. Such systems include HVACs and Clean utilities such as WFI (Water-for-Injection)

Diffusion blending

A process in which particles are reoriented in relation to one another when they are placed in random motion and interparticular friction is reduced as the result of bed expansion (usually within a rotating container). Also referred to as tumble blending.

Deviations

A deviation can be simply described as an unintended event which causes a test or verification to fail to meet expected acceptance criteria.

Degree of invasiveness

A device, which in whole or in part, penetrates inside the body either through a body orifice or through the skin surface, is invasive. Invasiveness is generally categorised as invasive of a body orifice (including the surface of the eye), surgically invasive devices and implantable devices.

Device Master Record (DMR)

a compilation of records containing the procedures and specification for a device. The contents of a DMR can contain local procedures such as SOPs and work instructions along with global or divisional specifications used to detail manufacturing processes, intermediate product or final product.

Drug Product

The dosage form in the final immediate packaging intended for marketing. The finished dosage form that contains a Drug Substance, generally, but not necessarily in association with other active or inactive ingredients. (FDA)

Duration of Contact

In determining the classification of a device the duration that the device is in continuous contact with the patient is defined as transient, short term or long term. The longer the device is in contact with the patient or user, the greater the risk and therefore this has to be taken into account when determining classification. Continuous use is defined in MEDDEV 2.4/1 as the uninterrupted actual use for the intended purpose. Where use of a device is discontinued in order that the device is immediately replaced with an identical device (e.g. replacement of a urethral catheter) this shall be considered as continuous use of the device.

E

Electronic Signatures

Electronic signatures are computer-generated character strings that count as the legal equivalent of a handwritten signature. The regulations for the use of electronic signatures are set out in 21 CFR Part 11 of the FDA. Each electronic signature must be assigned uniquely to one person and must not be used by any other person. It must be possible to confirm to the authorities that an electronic signature represents the legal equivalent of a handwritten signature. Electronic signatures can be biometrically based or the system can be set up without biometric features.

Encapsulation

The division of material into a hard gelatin capsule. Encapsulators should all have the following operating principles in common: rectification (orientation of the hard gelatin capsules), separation of capsule caps from bodies, dosing of fill material/formulation, rejoining of caps and bodies, and ejection of filled capsules.

Endotoxin

A pyrogenic product (e.g., lipopolysaccharide) present in the bacterial cell wall. Endotoxin can lead to reactions in patients receiving injections ranging from severe fever to death.

Equipment Qualification

Qualification means the process to demonstrate the ability to fulfil specified requirements. EQ consists of proving and documenting that equipment or ancillary systems are properly installed (Installation Qualification, IQ), work correctly (Operations Qualification OQ), and the different sub-systems work together as a system (Performance Qualification PQ) and actually lead to the expected results.

Qualification is part of validation, but the individual qualification steps alone do not constitute a validated process.

Excipient

Substances other than the API which have been appropriately evaluated for safety and are intentionally included in a drug delivery system to provide a specific role in manufacturing, shelf-life or physical property.

Equipment Range

The full range that equipment is capable of performing, as per the manufacturer specification and tolerances. (a process may not utilize the full equipment range, operating over a narrower range).

F

Factory Acceptance Testing (FAT)

An FAT or Factory Acceptance Test is an engineering activity that inspects and verifies that the equipment or system meets the requirements of the URS.

Failure Mode And Effects Analysis (FMEA):

A risk assessment tool that provides for an evaluation of potential failure modes and their likely effect on outcomes and/or product or process performance in order to prioritize risks and monitor the effectiveness of risk control activities. It is often used to identify areas within a given process, product, or system that render it vulnerable.

FDA 483s

An FDA 483 letter typically includes a summary of findings and observations in relation to an audit or inspection where the FDA representatives have reason to believe GMP or other regulations have been violated or are not being met. In response to an FDA 483 letter, the company should address each item and provide a timeline for correction or request clarification of what changes are required.

Functional Design Specification (FDS)

A functional design specification is a document that specifies how particular requirements are met – this can be a combination of how the equipment/process operates mechanically/automatically etc. An FDS is typically written to response to a URS

Fluid

A fluid is a substance that undergoes continuous deformation when subjected to a shearing force.

G

GAMP

Good Automated Manufacturing Practice (GAMP) is a set of guidelines for manufacturers and users of automated systems in regulated industries. Specifically, the Medical device, pharmaceutical and biopharmaceutical industries.

The application of GAMP and Validation of Automated Systems in manufacturing helps ensure that regulated medical devices and medicinal products have the required quality and are manufactured according to Good practices, meet regulatory and legal requirements and ensure patient safety.

Good Documentation Practices, GDP

The handling of written or pictorial information describing, defining, specifying and/or reporting of certifying activities, requirements, procedures or results in such a way as to ensure data integrit

Granulation

A process of creating granules. The powder morphology is modified through the use of either a liquid that causes particles to bind through capillary forces or dry compaction forces.

Grade A Areas

Aseptic processing areas, critical in nature where sterile products are exposed to the environment receiving no further sterilization. High-risk operations (for example aseptic stopperage, filling, loading of the lyophilizer) occur in Grade A areas. They are considered ISO 5 under both dynamic and static conditions.)

Grade B Areas

Aseptic processing areas where the sterile product is protected from the environment. Grade B processing areas are the background environments for Grade A areas and are considered ISO 7 environments in the dynamic state and ISO 5 environments under static conditions.

Grade C Areas

Non-critical areas where bulk product or materials are exposed to the environment, yet final sterilization has not yet been performed. Grade C areas are support areas for non-sterile production activities; purification, formulation, and preparation of components, equipment, etc. for sterilization. They are considered ISO 8 (Class 100,000) environments in the dynamic state and ISO 7 (Class 10,000) environments under static conditions.

Grade D Areas

Non-critical production areas, support areas, airlocks, or corridors. They are support areas for non-sterile production activities in closed systems; cell culture, or buffer and media preparation areas. Grade D Airlocks are used for the movement of product, materials, and personnel into classified areas.

GHTF

Global Harmonization Task Force

GxP

GxP is a general term for good practice with regard to quality guidelines and regulations. These guidelines are used in many fields, including the pharmaceutical, medical device and food industries. "x" is used as an umbrella letter representing different subjects or disciplines in industry. Some prime examples include GLP (Good Laboratory Practice), GDP (Good Documentation Practice), GEP (Good Engineering Practice) and GMP (Good Manufacturing Practices). Furthermore, the use of a lower case "c" as a prefix indicates "current" or "up-to-date"

H

Harm

Damage to health, including the damage that can occur from loss of product quality or availability.

High level risk assessment (HLRA)

A High level risk assessment that can be used at the beginning of a project to estimate the risk. Such as the risks involved with bringing in new computerised/automated equipment.

HVAC

Heating, ventilation and air-conditioning (HVAC) systems are used to control the environmental conditions within an area or manufacturing facility. HVAC systems also provide comfortable conditions for operators based in the manufacturing environment. Temperature, relative humidity (RH) and ventilation should not adversely affect the quality of products during their manufacture and storage, or the proper functioning of equipment

Hydrogel

A biomaterial made up of a network of polymer chains that are highly absorbent and as flexible as natural tissue.

☐

I

ICH

International Conference on Harmonization of Technical Requirements for Registration of Pharmaceuticals for Human Use.

Intended Purpose

Intended purpose means the use for which the device is intended according to the data supplied by the manufacturer on the labelling, in the instructions and/or in promotional materials. (Chapter I section 1 of Annex IX of Directive 93/42/EEC)

Impurity

Any component of the new active pharmaceutical ingredient which is not the chemical entity defined as the new active pharmaceutical ingredient OR any component present in the active pharmaceutical ingredient or final product which is not the desired product, a product-related substance, or excipient including buffer components.

Invasive device

A device, which, in whole or in part, penetrates inside the body, either through a body orifice or through the surface of the body.

IQ/OQ

Equipment IQ/OQ is defined as establishing documented evidence that all key aspects of the process equipment installation adhere to the manufacturer's approved specifications and any recommendations of the supplier of the equipment are suitably considered.
The process/equipment must also operate as intended and all user requirements are adequately fulfilled.

IFU

Instructions for Use.

(Plant) Injunction

An injunction is a judicial process initiated to stop or prevent violation of the law, such as to halt the flow of violative products in interstate commerce and to correct the conditions that caused the violation to occur. (FDA 21 U.S.C. 332; Rule 65, Rules of Civil Procedure).
If a firm has a history of violations and has promised correction in the past but has not made the corrections, the injunction is more likely to succeed. However, the freshness of the evidence is critical.

For an injunction action to be credible in the eyes of the Department of Justice (DOJ), the U.S. Attorney and the court, the evidence must be current. Timeliness is an important factor when considering an injunction action, with or without a Motion for Preliminary Injunction or a temporary restraining order (TRO). However, case quality and credibility must not be sacrificed to meet guideline time frames. The purpose of the guideline time frames is to limit, as much as can reasonably be expected, the need to update evidence. Updating entails extra work at all levels of the case development and review process and more importantly, delays obtaining an injunction which is intended to stop violations that adversely affect the safety or quality of products in commerce.

ISO

International Organization for Standardization. Agency responsible for developing international standards. E.g. ISO 13485 Medical Devices.

Isolator

A sealed enclosure, which provides full physical separation between the critical processing zone and the surrounding other processing zones. The internal surfaces of the isolator and of its contents are decontaminated, in accordance with defined objectives, by highly effective cycles. (e.g. Vaporised Hydrogen peroxide)

Enclosure capable of preventing ingress of contaminants by means of physical interior/exterior separation, and capable of being subject to reproducible interior bio-decontamination.

Isoelectric Precipitation

Isoelectric Precipitation works by reducing the electrostatic forces to near zero, allowing the proteins to precipitate out.

ISO 13485

ISO 13485, ISO standard, published in 2003, that represents the requirements for a comprehensive management system for the design and manufacture of medical devices.

ISO 14971

An ISO standard, published in 2007, that provides a framework and requirements for a risk management system for medical devices. This standard establishes the requirements for risk management to determine the safety of a medical device by the manufacturer during the product life cycle.

ISO 9001 - ISO 9001 is an ISO standard that represents the requirements for quality management systems. It is used across industries and is not specific to medical devices like ISO 13485.

Item Master

A of all components that a manufacturer buys, builds or assembles into its products. The item master includes information like the size, shape, material, manufacturer, manufacturer part number and vendor for each component.

IVD

In Vitro Diagnostic tests are medical devices intended to perform diagnoses from assays in a test tube, or more generally in a controlled environment outside a living organism.

IVDD

The In Vitro Diagnostic Device Directive delineates requirements that in vitro diagnostic devices must meet before they can be sold in the EU market.

Intermediate

A material produced during steps of the processing of an API that undergoes further molecular change(s) or purification before it becomes an API.

J

JIT (Just in time)

A strategy used to monitor inventory levels with the goal of reducing inventory and associated carrying costs.

K

Kanban

A scheduling system that advises manufacturers what to produce, when to produce and how much to produce. Pioneered by Toyota, the approach is based on demand. Inventory is replenished only when visual cues like an empty bin, trolley or cart show that it's needed.

L

Laminar flow

Laminar flow is when fluid particles move in parallel layers, at a constant velocity.

Lifecycle (Validation)

The Validation lifecycle refers to the requirement to control and document all validation activities from conception and URS stage to the retirement of equipment or a process. The lifecycle approach ensures compliance throughout the life of the process/equipment while maintaining a validated state throughout the application of change control.

Linearity

The ability of an analytical procedure (within a given range) to obtain test results that are directly proportional to the concentration (amount) of analyte in the sample.

Line Clearance

The act of performing and documenting the removal of materials from a production or packaging line and cleaning prior to the introduction of a new batch or lot.

Lyophilization (or Freeze Drying)

Lyophilization is the removal of ice or other frozen solvents from a material through the process of sublimation and the removal of bound water molecules through the process of desorption.

M

Maximum Allowable Carry Over (MACO)

The amount of allowed product residue (carry-over) from lot-to-lot, batch-to-batch. This limit is based on the most conservative or lowest level of three MACO calculation methods (1) Limited based on Toxicity, (2) Limit based on Smallest Therapeutic Dose, and (3) Worst Case Dose.

Measurement Capability Index (MCI)

The Measurement Capability Index (MCI) represents the capability of the measurement system. It is used to evaluate the capability of the gauge to classify product against predetermined specifications.

Measurement System Analysis (MSA)

A study to determine the degree of error involved in measuring the given parameter. The measurement system involves the combination of operations, procedures, gauges, instruments, environmental conditions, people and software.

Medical Device

A medical device is "an instrument, apparatus, implement, machine, contrivance, implant, in vitro reagent, or other similar or related article, including a component part, or accessory which is:

- recognized in the official National Formulary, or the United States Pharmacopoeia, or any supplement to them,
- intended for use in the diagnosis of disease or other conditions, or in the cure, mitigation, treatment, or prevention of disease, in man or other animals, or
- intended to affect the structure or any function of the body of man or other animals, and which does not achieve any of its primary intended purposes through chemical action within or on the body of man or other animals and which is not dependent upon being metabolized for the achievement of any of its primary intended purposes."

Medicinal Drug Products (Finished Products)

Finished dosage forms (e.g. tablet, capsule, or solution) that contain the active pharmaceutical ingredient usually combined with inactive ingredients. Medicinal products are intended to furnish pharmacological activity or other direct effect in the diagnosis, cure, mitigation, treatment, or prevention of disease or to affect the structure and function of the body.

MDD

The Medical Device Directive is intended to harmonize the laws relating to medical devices within the European Union. Medical Device Directive 93/42/EEC was most recently reviewed and amended by 2007/47/EC.

MHRA

The Medicines and Healthcare products regulatory Agency (MHRA) is the UK government agency which is responsible for ensuring that medicines and medical devices work and are acceptably safe.

MSDS

Material Safety Data Sheet.

N

NCR

Non-Conformance Report.

NIH

National Institutes of Health (U.S.)

Noel

No Observed Effect Level. In relation to Cleaning Validation.

Non-conformity

A deficiency in a characteristic, product specification, CQA, process parameter, record, or procedure that renders the quality of a product unacceptable, indeterminate, or not according to specified requirements.

Non Parametric Data

Where the type of data is non variable Also referred to as attribute data eg (Visual inspection resulting in a PASS/FAIL result.

Notified Bodies

A notified body is a certification organisation which the national authority (the competent authority) of a member state designates to carry out one or more of the conformity assessment procedures or audits described in the annexes of the medical devices directives or GMP legislation.

NPI (New product introduction)
The market launch or commercialization of a new product. NPI takes place at the end of a successful product development project.

O

Open System

An environment in which system access is not controlled by persons who are responsible for the content of electronic records on the system (21 CFR, Part 11)

Outlier

A test result that is statistically different compared to a set of other test results obtained from the same sample or samples from the same lot of material.

Out-Of-Specification

A recorded result that falls outside the established specification(s) or acceptance criteria.

Out-Of-Trend

Analytical result, which is within specification or acceptance criteria, but different from those usually obtained or expected. Out-of-trend results should be investigated by the same general principles as out-of-specification results.

Quantitation limit

The lowest amount of analyte in a sample which can be quantitatively determined with suitable precision and accuracy for an analytical procedure. The quantitation limit is a parameter of quantitative assays for low levels of compounds in sample matrices and is used particularly for the determination of impurities and degradation products.

Overall Equipment Effectiveness(OEE)

A calculation for measuring the efficiency and effectiveness of a process, by Equipment breaking it down into three constituent components (the OEE Factors) Availability x Performance x Quality.

Overkill

Sterilization process that is demonstrated as delivering at least a 12 Spore Log Reduction (SLR) to a biological indicator having a resistance equal to or greater than the bioburden level.

P

Pan Coating

The uniform deposition of coating material onto the surface of a solid dosage form while being translated via a rotating vessel.

Particle count test

Test covers verification of cleanliness. Dust particle counts measured. The number of readings and positions of tests should be defined in accordance with ISO 14644-1 Annex B5.

Performance indicators

Measurable values used to quantify quality objectives to reflect the performance of an organization, process or system, also known as performance metrics in some regions. (ICH Q10)

Performance Qualification (PQ)

Establishing by documented evidence that the process, under anticipated (controlled) conditions, consistently produces a product which meets predetermined requirements.

Precision

The degree of agreement (scatter) between a series of measurements when a method is applied repeatedly to multiple samplings of a homogeneous sample or artificially prepared sample under the prescribed conditions. There are three types of precision; repeatability, intermediate precision and reproducibility.

Pressure cascade

A process whereby air flows from one area, which is maintained at a higher pressure, to another area at a lower pressure.

Piping & Instrument Diagrams (P&IDs)
Engineering technical drawings that provide details of the connections and integration of equipment, services, material flows, plant controls and alarms. The P&ID also provide the reference for each tag or label used for identification.

PMA

Premarket approval by FDA is the required process of scientific review to guarantee safety and effectiveness for Class III devices.

PMDA

The Pharmaceutical and Medical Devices Agency in Japan reviews applications for marketing approval of pharmaceuticals and medical devices. It also monitors their post-marketing safety and provides relief compensation for people who have suffered from adverse drug reactions from pharmaceuticals or infections from biological products.

PMS

Post Marketing Surveillance is the practice of monitoring a pharmaceutical drug or device after it has been released on the market.

Process design

Defining the commercial manufacturing process based on knowledge gained through development and scale-up activities.

Process qualification

Confirming that the manufacturing process as designed is capable of reproducible commercial manufacturing.

Process window

The selected operating range of machine setting/parameter that will produce product to meet all quality and product specifications.

Product Recovery

Product recovery is a critical and important step in the process. It is also referred to as "Downstream processing". It is often the most expensive step in the process. For recombinant-DNA derived products, purification can often account for 90% of the total production costs.

Prospective Validation

Prospective Validation is when validation is done in advance of commercial manufacturing.

Procedures

Also known as Standard Operating Procedures, or SOPs, give directions for performing certain operations.

Protocols

Give instructions for performing and recording certain discreet operations. (Examples include engineering protocols, validation protocols etc.)

Pure

A term typically used within pharmaceutical manufacturing, a product or substance is pure if it is free of contaminants, foreign matter, chemicals and harmful microbes.

Q

QMS

Quality Management System can be expressed as the organizational structure, procedures, processes and resources needed to implement quality management.

Quality

The degree to which a set of inherent properties of a product, system, or process fulfils requirements. (ICH Q9)

Quality by design

This is a systematic approach that begins with predefined objectives and emphasizes product and process understanding and process control, based on sound Science and engineering principles.

Quality Management System

A Quality Management System, often abbreviated to (QMS) is any system based on a collection of business processes that are primarily focused on providing safe and quality products that consistently meet customer requirements.

☐
Quarantine

The status of materials isolated physically or by other effective means pending a decision on their subsequent approval or rejection.

(Quality) Policy

A document in which a company or organization outlines their commitment and approach to quality. It usually sets out how they plan to achieve a high and consistent standard of quality. It should in some way speak to the customer or end user.

Qualification Plan

A Qualification Plan (QP) describes all the qualification measures and at which stage of the qualification the verification will be completed. It typically contains detailed descriptions of the necessary test measures and a description of the interdependencies of the individual tests. In some instances, there may not be a need or a requirement for a qualification plan. A validation plan can also serve to detail the qualification strategy.

QP

Companies that intend to manufacture or import medicinal products or intermediate products, for use in clinical trials or for market within the EU, must appoint the service of a Qualified Person, in order to comply with EU Good Manufacturing Practice Standards.

QPM

Quality Policy Manual.

QSP

Quality System Procedure.

QSR

Quality System Regulations.

R

Range

Range is defined as the interval between the upper and lower measurements required. The minimum specified range should be within the equipment range and validated to operate at all points within the range.

Recall

As defined at 21 CFR 7.3(g), "recall means a firm's removal or correction of a marketed product that the Food and Drug Administration considers to be in violation of the laws it administers and against which the agency would initiate legal action, 2 21 CFR 806.2(h). e.g., seizure. Recall does not include a market withdrawal or a stock recovery." Recall does not include routine servicing. Recall also does not include an enhancement, as defined by this guidance.

Relative humidity

The ratio of the actual water vapour pressure of the air to the saturated water vapour pressure of the air at the same temperature expressed as a percentage. More simply put, it is the ratio of the mass of moisture in the air, relative to the mass at 100% moisture saturation, at a given temperature.

Reusable medical device

A device intended for repeated use either on the same or different patients, with appropriate decontamination and other reprocessing prior to re-use.

Reusable Surgical Instrument

Instrument intended for surgical use by cutting, drilling, sawing, scratching, scraping, clamping, retracting, clipping or similar surgical procedures, without connection to any active medical device and which are intended by the manufacturer to be reused after appropriate procedures for cleaning and/or sterilisation have been carried out.

Re-Qualification

Requalification is designed to verify and ensure that the equipment/instrument/system is maintained in a qualified state after modification or after a stipulated time period (downtime).

Residual Risk

The risk level remaining after applying the identified controls on a high risk of harms and hazards manifestation.

Resolution

The smallest change in quantity that can be detected or provided by an instrument.

Residual Solvent

Organic volatile chemicals used or produced during the manufacture of APIs or excipients, or in the preparation of medicinal products.

Retain Samples

Samples that are kept for potential investigations and retests. It should be noted that retain samples are not a regulatory requirement, per Annex 10 or 21 CFR part 11.

Retrospective Validation

Retrospective validation is used for facilities or processes that have not completed formal validation. Historical data or a retrospective review can provide the evidence that the process or facility is operated as intended.

Rinse Sampling

Using a solvent to contact all surfaces of the sampled item to quantitatively remove target residue. The solvent can be water, water with pH adjusted, or organic solvent.

Right First Time

Right First Time strives to create a culture of excellence. People are challenged with performing their tasks always in the correct manner to achieve the correct results always - right the first time.

Risk

The combination of the probability of occurrence of harm and the severity of that harm.

Risk Management

Risk management involves the systematic application of management policies, practices and procedures that identify, analyse, control and monitor risk.
It is important to recognise that risk management should begin at the outset of the design and development phase of a project. The first step is to identify the user needs and intended use and application of the device.

RoHS

"Restriction of Hazardous Substances in electrical and electronic equipment 2002/95/EC". An initiative that was adopted by the European Union (EU) in February 2003 and put into effect July 1, 2006
Ruggedness

An indication of how resistant a test method or process is to typical variations in operation, such as those to be expected when using different analysts, different instruments and different reagent batches.

☐

S

Scaffold

A structure of artificial or natural materials on which tissue is grown to mimic a biological process outside the body.

SKU

(Stock keeping unit) A unique sales stock identifier.

Specifications

A approved document detailing the requirements with which the products or materials used or obtained during manufacture have to conform. They serve as a basis for quality evaluation.

Specificity

The ability to assess unequivocally the analyte in the presence of components, which may be expected to be present.

Stability

Stability studies are used to demonstrate and justify assigned expiration or retest dates.

5S

5S is a Japanese methodology of organising and storing items in a work or lab environment. It has been adopted by many Western companies as a tool to help maintain standards and reduce errors and mix-ups. The "5s" represents each stage of the method.

Sort

Sorting out any items that are not in use and removing to a more appropriate area or to storage or the bin.

Set-in-Order

The idea of "Set-in-Order" is to be always organised. "A place for everything and everything in its place. "If we "set-in-order" we can help to make live processing and testing more efficient and reduce the risk of errors, omissions and accidents.

Shine

Regular cleaning is an important practice and it is always helpful to "Clean as you go."

Standardise

Implement standard practices through SOPs and training. Standardisation can also be applied to work station layout.

Sustain

Make it a habit! After implementing a 5s methodology, it is only effective if continuous efforts are made to "sustain" the changes.

Sterility Assurance (SAL)

Probability of a single viable microorganism occurring on an item after sterilization. For a terminally sterilized medical device to be designated as "sterile", the minimum sterility assurance level shall be SAL = 10-6 or better. When applying this quantitative value to assurance of sterility, an SAL of 10-7 has a lower value but provides a greater assurance of sterility than an SAL of 10-6 .

T

Tableting

The reconstitution of a powder blend in which compression force is applied to form a single unit dose. (tablet)

Tableting press

Tablet press subclasses primarily are distinguished from one another by the method that the powder blend is delivered to the die cavity. Tablet presses can deliver powders without mechanical assistance (gravity), with mechanical assistance (automation), by rotational forces (centrifugal), and in two different locations where a tablet core is formed and subsequently an outer layer of coating material is applied (compression coating).

Traceability Matrix

A Traceability Matrix is a document that links the user requirements and specifications to where the verification and testing has been documented within the validation activities.
A traceability matrix illustrates that all user requirements are traceable to the evidence based test.

Turbulent flow

Turbulent flow is when the movement of fluid particles are varying in velocity and direction.

U

Uniform

The product is manufactured consistently and will have the same quality between batches manufactured on different days.

UDI, Unique Device Identification

The UDI is a series of numeric or alphanumeric characters that is created through a globally accepted device identification and coding standard. It allows the unambiguous identification of a specific medical device on the market.

Uninterrupted Power Supply

An uninterruptible power supply (UPS) is a system for buffering the main power supply. If the power supply fails, the battery of the UPS supplies the required power. When the power supply returns, the UPS battery stops supplying power and is recharged.

Unit Operation

Unit operations are the individual steps in the process that modify materials and their properties at each step of the process. Each unit operation comes together to create a complete process.

User Requirement Specification, URS

The URS is a critical document that defines the requirements of a particular system, equipment or process. Requirements such as the functional and operational aspects of the system are typically documented here.

USP

United States Pharmacopoeia.

V

Validation

Validation is confirmation via documented evidence that the particular requirements for a specific intended use can be consistently fulfilled under anticipated conditions.

Validation Master Plan

A document providing information on a company's validation work programme. It typically details timescales for the validation work to be performed along with the key deliverables.

Verification

Verification confirmation by examination and provision of objective evidence (i.e. documentation) that the specified requirements have been fulfilled.

Vaporized Hydrogen Peroxide (VHP)

Vaporization of liquid hydrogen peroxide which results in a mixture of VHP and water vapor. The VHP mixture is used to decontaminate isolators.

W

Warning Letter

A warning letter is a correspondence that notifies regulated industry about violations that FDA has documented during its inspections or investigations.

WEEE Directive

Waste electrical and electronic equipment directive. European Community directive 2002/96/EC where manufacturers are responsible for disposing of electrical/electronic waste.

WFI (Water for injection)

WFI is sterile and pyrogen free water containing o less than 10 CFU/100ml (Colony Forming Units) with a sample size of between 100 and 300 ml and an endotoxin level < 0.25 EU/ml.

WHO

World Health Organization.

WI

Work Instructions.

Witnessed By

When signed or initialed is legal proof that the individual signing is physically present and observes the step, calculation, or operation being performed by someone else, and that all entries of data are true and accurate.

Worst Case

A set of conditions or parameters which, in combination with product specification or attributes at their limits, pose the greatest challenge to the process.

X

--

Y

--

Z

Zone Classification

Zone classification refers to GMP areas which include controlled (aka classified) and non-controlled manufacturing areas. Areas may be classified based on EU Grades A–D and/or ISO Class 5–8 (in the US - Class 100–Class 100,000 areas.